Dear Parents and Educators,

Welcome to Penguin Young Readers! As parents and educators, you know that each child develops at their own pace—in terms of speech, critical thinking, and, of course, reading. Penguin Young Readers recognizes this fact. As a result, each Penguin Young Readers book is assigned a traditional easy-to-read level (1–4) as well as an F&P Text Level (A–R). Both of these systems will help you choose the right book for your child. Please refer to the back of each book for specific leveling information. Penguin Young Readers features esteemed authors and illustrators, stories about favorite characters, fascinating nonfiction, and more!

Animal Allies
Creatures Working Together

LEVEL **4**

F&P TEXT LEVEL **R**

This book is perfect for a **Fluent Reader** who:
- can read the text quickly with minimal effort;
- has good comprehension skills;
- can self-correct (can recognize when something doesn't sound right); and
- can read aloud smoothly and with expression.

Here are some **activities** you can do during and after reading this book:
- Problem/Solution: Many of the animal ally pairings in this book help each other solve problems/find solutions. Have you had a friend help you to solve a problem? Have you found a solution to a friend's problem?
- Make Connections: Do you have any pets that you consider a close friend or ally? Think about how that animal came into your life and how you became close.

Remember, sharing the love of reading with a child is the best gift you can give!

D1132386

For Kim Whaley, my friend and
partner on so many projects—thanks
for making working and playing
together creative and fun!—GLC

PENGUIN YOUNG READERS
An imprint of Penguin Random House LLC, New York

First published in the United States of America by Penguin Young Readers,
an imprint of Penguin Random House LLC, New York, 2023

Photo credits: used throughout: (photo frame) happyfoto/E+/Getty Images; cover, 3: BirdImages/
E+/Getty Images; 4–5: (background) John Pitcher/Design Pics/Getty Images; 5: JohnPitcher/iStock;
6: Cannasue/iStock/Getty Images; 7: (top) Andrew Holt/Getty Images, (bottom) Edwin Remsberg/
The Image Bank/Getty Images; 8: Willie van Schalkwyk/Moment/Getty Images; 9: Vicki Jauron,
Babylon and Beyond Photography/Moment/Getty Images; 10: David Tipling/Education Images/
Universal Images Group/Getty Images; 11: James Keith/Moment/Getty Images; 12: Artush/iStock/
Getty Images; 13: Sixtine Derville/iStock/Getty Images; 14: (top) Nigel Dennis/imageBROKER/Alamy
Stock Photo, (bottom) Steve Adams/iStock/Getty Images; 15: FLPA/Alamy Stock Photo; 16: Jakob
Ziegler/iStock/Getty Images; 17: FionaAyerst/iStock/Getty Images; 18: Westend61/Getty Images;
19: slowmotiongli/iStock/Getty Images; 20: Paul Cowell/EyeEm/Getty Images; 21: oksanavg/iStock/
Getty Images; 22: EcoPic/iStock/Getty Images; 23: tatisol/iStock/Getty Images; 24: simonkr/E+/
Getty Images; 25: kitz-travellers/iStock/Getty Images; 26: Puneet Vikram Singh/Moment Open/Getty
Images; 27: Mark Horton/500Px Plus/Getty Images; 28: phototrip/iStock; 29: David bm/iStock/Getty
Images; 30–31: Oxford Scientific/Photodisc/Getty Images; 32: KATERYNA KON/Science Photo Library/
Getty Images; 33: Mohamed Tazi Cherti/500px/Getty Images; 34: Rob Jansen/iStock/Getty Images;
35: Christian Ziegler/Danita Delimont/Alamy Stock Photo; 36: (top) Troy Harrison/ Moment/Getty
Images, (bottom) milehightraveler/iStock/Getty Images; 37: Thomas Kline/Design Pics/Getty Images;
38: Shellphoto/iStock/Getty Images; 39: (top) Pete Oxford/Nature Picture Library/Alamy Stock Photo,
(bottom) Nature Picture Library/Alamy Stock Photo; 40: Jens Otte/iStock/Getty Images; 41: FLPA/
Alamy Stock Photo; 42: kobuspeche/iStock/Getty Images; 43: (top) AOosthuizen/iStock/Getty Images,
(bottom) Herbert Kratky/imageBROKER/Getty Images; 44–45: Mathieu Meur/Stocktrek Images/Getty
Images; 46: (top) John Howard/DigitalVision/Getty Images, (bottom) StefaNikolic/E+/Getty Images;
47: Halfpoint/iStock/Getty Images; 48: wundervisuals/E+/Getty Images

Visit us online at penguinrandomhouse.com.

Library of Congress Cataloging-in-Publication Data is available.

Manufactured in China

ISBN 9780593521915 (pbk)   10 9 8 7 6 5 4 3 2 1 WKT
ISBN 9780593521922 (hc)   10 9 8 7 6 5 4 3 2 1 WKT

Animal Allies

Creatures Working Together

by Ginjer L. Clarke

Introduction

Many animals help other animals.
Some keep each other safe from danger.
Others offer cleaning services. Some
share their homes. The coyote and
American badger even hunt together,
although they compete for food.

All of these animals are allies. They combine their skills, so they both live better together. This teamwork is called **symbiosis** (say: sim-bye-OH-sis).

The reasons why creatures work together can be surprising. Let's find out which animals are allies!

Bird Besties

Most birds stay away from animals with big mouths and sharp teeth. But the little Egyptian plover is not afraid. Some people say it jumps *inside* a Nile crocodile's mouth!

The crocodile gets bits of food stuck in its teeth. It opens wide when it wants help getting rid of the food. *Pluck!* The brave plover hops in to clean the croc's teeth— and to get a snack. The croc does not eat the bird, as thanks for its help.

The best animal partners both give and get help. This type of symbiosis is called **mutualism** (say: MEW-choo-ul-is-um) because

their actions are mutual—for each other. Everyone wins!

Another bird providing cleaning services is the oxpecker. It is also called the "tickbird." Why? It pecks bloodsucking ticks off large African animals.

The oxpecker uses its beak to pick at a giraffe's neck. It eats the ticks it finds and drinks the blood inside the ticks. It even pecks inside a zebra's ear.

The giraffe falls asleep while being cleaned. *Shriek!* The oxpecker calls loudly and flaps its wings to wake up the giraffe. It warns the giraffe of danger. This bird bestie helps its friend stay clean and safe.

The cattle egret also has big buddies.

Stomp! A rhinoceros plods along. Lots of grasshoppers hiding in the grass suddenly fly into the air. They are trying to flee the rhino's big feet. *Chomp!* The egret snaps up the escaping insects.

In return, the egret plays lookout because the rhino has poor eyesight. The egret sits on the rhino's back. It makes a lot of noise when it sees a predator, so the rhino moves somewhere safer. What a feathered friend!

Unlike rhinos, ostriches have good eyesight. Their long necks let them see very far. But they could miss a sneak attack when they lower their necks to eat grass. Zebras cannot see well, but they do have excellent hearing. So they team up!

Ostriches and zebras travel in groups. They take turns eating, so one is always keeping watch. A lion comes near the group. The zebras hear it first and send a warning. *Bray, bray!* Get away! The ostriches cannot fly, but they do run fast.

A much smaller African bird is the honeyguide. It is homies with the honey badger, one of the toughest animals around. These two are named for their favorite food. What is it? Honey, of course! But they cannot get honey without working together.

The smart honeyguide finds beehives by following bees to their trees. But the little bird is not strong enough to break the hive. *Tweet, tweet!* The bird alerts the honey badger and leads it to the hive. The honey badger rips open the hive and slurps out honey. Then the honeyguide eats the honeycomb. *Sweet!*

Water Workers

Sharks may seem scary, but they can be good allies. The tiger shark gives a free ride to a remora fish. This small fish is nicknamed the "sharksucker." *Slurp!* It has a flat head and a special disk that sucks onto the shark.

Sharks are messy eaters. The remora gets free food as it eats the leftovers from the shark's meal. It then keeps the shark's skin clean. Nice work!

Remoras also suck onto other kinds of sharks and some whales, sea turtles, and large stingrays. Sometimes they even attach to the skin of a human diver!

The cleaner wrasse (say: RASS) assists lots of sea creatures. How? It cleans their teeth! *Wiggle, wiggle!* The wrasse does a dance to show that it is ready to work.

Open wide! A moray eel stops moving and opens its mouth. The wrasse swims inside. *Nibble!* It picks out food from the eel's teeth and gills. The eel could easily eat the fish, but it does not hurt its helper.

The wrasse's customers line up to wait their turn. The wrasse works fast. It can clean more than 300 big fish in an hour!

A sea anemone (say: uh-NEM-oh-nee) looks like a plant, but it is really a simple creature. An anemone stings fish and eats them. But not the clownfish! This colorful fish can safely hide in a sea anemone. It is not harmed by the poison from the anemone's tentacles. And it helps the anemone get food.

The clownfish's bright colors make another fish swim closer. *Bam!* The fish gets stung. *Wham!* The anemone sucks the fish into its tube-shaped mouth. The clownfish cleans up any leftover bits.

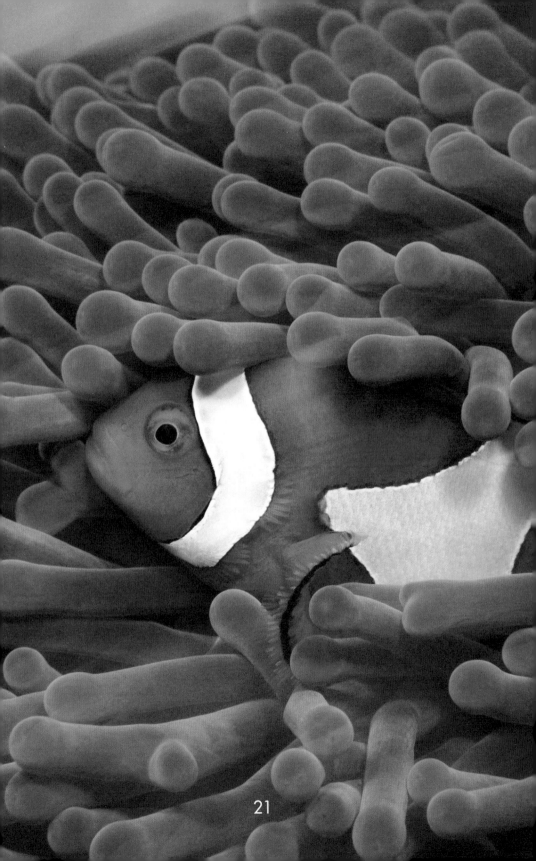

The hippopotamus has lots of helping partners. It spends most of its time in the water. The hippo's skin gets covered in insects and algae (say: AL-jee)—tiny water plants. It cannot remove these pests by itself.

Here comes the hippo-cleaning team! An African helmeted turtle climbs onto the hippo's back. *Munch!* It eats the plants and suns itself. Then an oxpecker lands and picks up some tasty ticks. *Crunch!*

The final members of the cleanup crew are labeo (say: LAH-bee-oh) fish. They swim all over the hippo and inside its huge mouth to clean its skin, teeth, and tongue. What a helpful bunch!

The marine iguana is the only lizard in the world that lives in the ocean. The only food it eats is algae. The iguana gets a lot of bugs on its body from living in the water. Its cleaner is the colorful Sally Lightfoot crab.

The iguana suns itself on a rock. *Skitter!* A big red crab scrambles all over the iguana. *Scatter!* The crab quickly picks ticks and dead skin off the iguana and runs away. The crab eats the ticks and skin as a tasty snack once it is safe.

The iguana and crab are more like frenemies, though. Sometimes the huge

crab steals and eats a baby iguana. *Oh my!*

Mini Mates

Aphids (say: A-fids) are itty-bitty bugs that live on and harm plants. But black ants love aphids for food. No, the ants do not eat the aphids! They get honeydew from them.

Aphids suck sap from plants. The sugar becomes honeydew in their bodies. *Tap! Tap!* The ants touch the aphids on their backs to release the honeydew.

The ants drink the honeydew and store any extra in their nests. In return, they take good care of the aphids. *Whap!* They even attack ladybugs that try to catch and eat the aphids.

Tree ants and rufous woodpeckers are strange partners because they are sometimes enemies. The woodpecker usually eats lots of tree ants. *Pow!* The tree ants sting the woodpecker. They protect their nest hanging in a tree.

Then something surprising happens in the spring. The ants make a nest. They leave a hole where a female rufous woodpecker can lay her eggs. *Wow!*

The ants leave the woodpecker chicks alone after they hatch. The mother bird does not eat the ants, as her thanks for using their nest. She watches over the nest until the chicks leave. Then the bird and the ants are enemies again—until next spring!

Some creatures help others but do not get anything in return. Helping does not harm them, so they do not mind. This is called **commensalism** (say: cuh-MEN-sul-is-um). *Commensal* means "eating together."

The pseudoscorpion (say: SOO-doh-SKOR-pee-un)—or "fake" scorpion—is a tiny creature. It looks like a scorpion with no stinger. It cannot fly or even move quickly. So it catches a ride!

A fly lands nearby. A pseudoscorpion gently grabs onto the fly using its pincers. *Whee!* It is safe from predators while it is on the fly. It drops off after it gets where it wants to go. It is home free!

31

Bacteria are super-small living things. They are everywhere, even in your body. *Eww!* But some bacteria are useful to the creatures they call home. They can help you and other animals digest food.

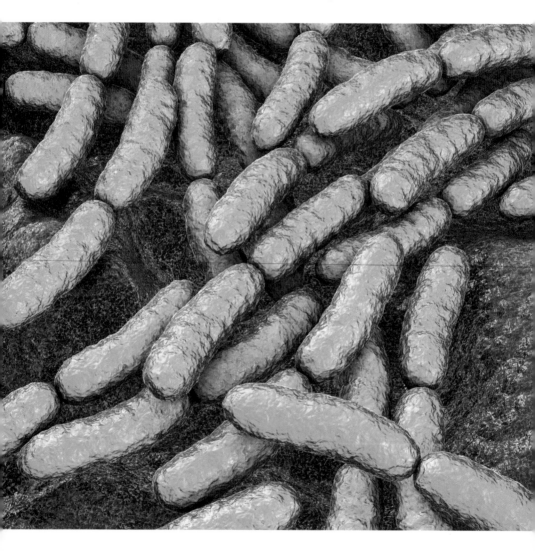

The giant panda usually eats only one thing—a plant called bamboo. *Chew, chew!* The panda eats so much bamboo that it needs aid to digest it. That is where the bacteria come in! Good bacteria in the panda's stomach keep it healthy and happy.

A sloth eats only plants. It also has plants living inside its fur. The sloth moves *soooo* slowly that its fur is covered in green algae. The algae make the sloth blend in with the trees. The tiny plants also provide some helpful food for the sloth.

The sloth has bug buddies, too. Sloth moths live only in the sloth's fur. The adult moths eat the algae. But the baby moths eat something special—sloth poop! *Yuck!*

Helpful Homes

The beaver is an important animal partner. It helps many river creatures by changing the area where they all live.

Chomp, chomp! A beaver bites through small trees until they fall down. Then it moves the trees to make a dam that turns a fast-moving river into a pond. It also digs tunnels underwater. The beaver is a hard worker.

Now salmon and frogs can lay their eggs in the calm water. Swans and ducks build nests on the dam. And otters have their babies in the tunnels. They all live in the beaver pond together. What a happy home!

The gopher tortoise makes a home for other animals, too. It gets its name from the gopher, a large rodent that burrows underground. *Dig, dig!* The gopher tortoise creates a big burrow in the sand. It can be about 25 feet long and 10 feet deep.

More than 300 different kinds of animals use the tortoise's home. The gopher frog and the eastern indigo snake both live there in summer to escape the heat. The burrow is also one of the only safe places for animals to hide during a forest fire.

The lesserblack tarantula also makes a burrow. It sleeps in the burrow during the day. At night, it hunts insects and frogs. But the spider lets one animal share its burrow—a tiny frog!

The dotted humming frog seems to be this spider's friend. They even lay their eggs side by side. How cute!

Scoot, scoot! The teeny frog hides under the big spider for safety from snakes. The frog also feeds the spider's leftovers to its tadpoles. The frog aids the spider by munching ants that eat the spider's eggs.

Termites build a huge dirt home called a mound. They leave the mound when their queen dies. The mound then becomes a home for many other animal friends. First, a family of dwarf mongooses moves in.

But mongooses are messy. So they share the mound with a plated lizard. The lizard keeps their home neat and tidy. How? It eats the mongooses' poop!

The dwarf mongooses work with a hornbill bird, too. They look for insects together. The bird catches bugs that the mongooses dig up. The bird also cries out if it sees danger. *Squawk!* Run, do not walk! The mongooses hurry back to the mound.

The goby is a small fish that lives on
the seafloor. It cannot dig, so it shares a
home with a blind shrimp. The shrimp
has big claws for digging a hole, but it
cannot see. So the goby keeps watch.

Eeek! The goby swims out first if all is quiet. The shrimp follows and touches the fish's tail fin with its feelers. *Eeek!* The goby sees a shark. It flicks its tail to signal to the shrimp. They both swim back inside.

The shrimp seals off the hole with sand. They sleep safely inside together.

Who is the best animal ally? Dogs!
They are many humans' best friends.

Dogs have lived
with humans for
thousands of years.
They stayed with
people who gave
them food and
protected the people.

Arf, arf! Guard dogs act like burglar alarms. They bark if they see, hear, or smell danger. This warns their people and keeps away any trouble.

Dogs do a lot of other jobs, too. They pull sleds, hunt birds, herd sheep, and guide people who cannot see well. Many dogs are great at giving cuddles and kisses. *Aw!*

Some animals get by with a little help from their friends. Even animals that could hurt or eat each other can get along for a while. Many creatures live best by working together. Can you be an ally, too?